FINAL ACCEPTANCE TESTING

FINAL ACCEPTANCE TESTING

Michael J. Muolo

ASQC Quality Press
Milwaukee, Wisconsin

FINAL ACCEPTANCE TESTING
Michael J. Muolo

Library of Congress Cataloging-in-Publication Data

Muolo, Michael J.
Final acceptance testing / Michael J. Muolo.
p. cm.
Includes bibliographical references and index.
ISBN 0-87389-126-0
1. Reliability (Engineering) I. Title.
TA169.M86 1992
620'.00452 — dc20 91-24399
CIP

10987654321

ISBN 0-87389-126-0

Acquisitions Editor: Jeanine L. Lau
Production Editor: Mary Beth Nilles
Set in Eras by DanTon Typographers. Cover design by Dick Skover.
Printed and bound by BookCrafters.

Printed in the United States of America

ASQC Quality Press
611 East Wisconsin Avenue
Milwaukee, Wisconsin 53202

To

Robin, Michael Jin, and Melissa Kim

and

The System Integration Office (SIO)
''Integrity from Earth to Stars''

CONTENTS

PREFACE

The goal of this book is to help managers understand, organize, develop, and execute final acceptance testing. The process, which assumes that operational test and evaluation (OT&E) has been completed, gives the user performance feedback about final system acceptance. This testing method can be applied over a wide spectrum of systems and test situations, providing an entire system evaluation. This book addresses personnel needs, systems to be tested, test methods, data analysis, and performance reporting. More specifically it covers the determination of system usability, in a dedicated operational environment, including operational functionality and technical reliability.

My work involves designing and managing annual performance tests of ballistic missile, atmospheric, and space tactical warning and assessment systems. This book is a composite of my ideas and experiences, along with references that support this test method.

This book does not discuss the performance limitations or capabilities of the early warning system, nor does it reflect the established policies of the Department of Defense (DOD). This book was not produced on government time or with government resources. It was submitted and approved for release by the responsible government agency on November 5, 1990, prior to publication. Releases for referenced testing documents, diagrams, and regulations have been obtained where appropriate; some material is in the public domain.

Chapter 1.

INTRODUCTION

This book explains the role of management in final acceptance testing. Final acceptance testing is done after operational test and evaluation (OT&E). It provides performance feedback about system acceptance. There are many applications for this test method, which emphasizes the importance of testing an entire system from stimulus input to final system output, including peripheral equipment. This type of performance testing is a reliable method for identifing both minor and serious system deficiencies, preserving the system life cycle, ensuring proper implementation of changes, and guaranteeing proper system output.

1.1. Background.

Whether they're new or being redesigned, all systems are designed and developed to comply with a design specification and a specific need. The quality of each component in the system contributes to the overall system performance. To ensure that the system remains within specification tolerance, each subsystem must be tested. During the developmental stage, the first testing that a new or changed system undergoes is called developmental test and evaluation (DT&E).

DT&E testing is, for the most part, performed off-line. It's safer, cheaper, easier to schedule, more flexible, and doesn't tie up operational resources (equipment and people) as on-line activities do. However, DT&E testing does not satisfy the specification for independent processing and functioning within the system's actual operating environment. To ensure full system development and performance, DT&E must be supplemented by operational test and evaluation (OT&E).

OT&E is necessary to guarantee that the new or changed system meets the user's requirements, which include operational needs and maintenance plans. OT&E provides an opportunity for an objective evaluation of the system's operational capabilities, limitations, and needed improvements. Users or operators consider the results of these operational tests when deciding on system deployment.

Final acceptance testing, which is done after DT&E and OT&E, concentrates on the system's suitability for either the end user or the intended operation. The test parameters include design, coordination, execution, analysis, and reporting. Final acceptance testing allows management to answer one critical question: is the system under evaluation technically and operationally suitable?

Some might argue that since OT&E testing is performed in an operational environment, further testing is not required. But OT&E might not completely satisfy all evaluation needs or circumstances, for example with products such as computers, motorized vehicles or equipment, and various types of electronic equipment. A consumer who is shopping for a car usually takes a test drive to determine the car's operability and suitability. The consumer is in effect conducting an evaluation process by collecting data and information and then making a comparative analysis against other automobiles. The result is a product selection that satisfies the buyer's overall needs and desires. Thus, final acceptance testing is a simple but scientific testing tool for management to determine system compatibility, suitability, and operability.

Figure 1.1, "Managerial Process for Final Acceptance Testing," depicts the entire managerial process that forms the basis of this book. Selected chapters contain application examples that managers can apply to their own test situations. The appendix includes sample test planning checklists, information management forms, and examples of test plans and procedures corresponding to each example.

1.2. Application Example.

To demonstrate the usefulness and versatility of this test process, this book will follow an example involving a hypothetical surveillance radar detection system. This radar system is completely assembled and tested by the manufacturer. It is representative of a ready-to-install radar with specific system capabilities and limitations. System upgrades and enhancements can be made available once the initial radar system is installed. The following system specification applies.

Purpose. This surveillance radar detection system is used to detect and track aircraft for small independently operated airports.

Capabilities. Track a maximum of 20 aircraft 120 nautical miles from azimuth center and provide 360 degrees of earth horizon-to-horizon coverage.

General.

System Designation: RDS-120 Tracking System

Manufacturer/Developer: Radar-Systems, Inc.

Prospective Owner/Operator: Rocky Mountain Airport

Airport Description: This is a small airport handling private single engine aircraft within the Colorado Rocky Mountain Range. Maximum daily takeoff and landing traffic load is 15 aircraft. Landing strip also serves as emergency runway for aircraft encountering severe mountain weather conditions.

Radar-Systems, Inc. has successfully completed the required DT&E and OT&E testing for the RDS-120. The owner and manager of Rocky Mountain Airport is investigating various radar systems to replace the existing system, which is outdated. Before accepting the system, Rocky Mountain Airport wants to test it to ensure its usability. More specifically, this will be a final acceptance test prior to system acquisition. The airport's owner and manager has requested that the testing be done by the system operators who will actually use the system. In this example, the test environment will involve a Radar-Systems, Inc. test facility and equipment located at another small airport. Radar-Systems, Inc. operators will be on hand to instruct the Rocky Mountain Airport radar operators. To ensure that the test will exercise all user requirements, the airport owner and manager will develop test requirements according to the process that is shown in Figure 1.1 and discussed step-by-step in the following chapters.

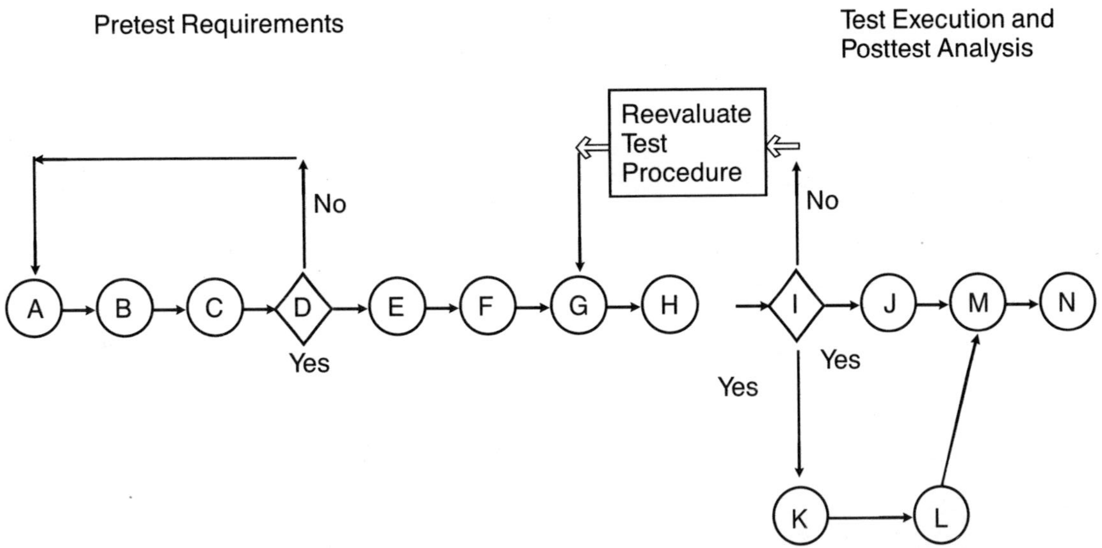

A. Test Requirements and Purpose
B. Test Objectives
C. Test Capabilities
D. Is the test feasible?
E. Designing a Test
F. Developing a Test Plan
G. Developing a Test Procedure

H. Test Execution
I. Did the test execute?
J. Quick-Look Test Evaluation
K. Data Collection Requirements
L. Analysis of Test Data
M. The Test Report
N. Lessons Learned

Figure 1.1: Managerial Process for Final Acceptance Testing. This program flow chart depicts the pretest coordination and test execution/posttest data analysis process. This diagram is a composite of my testing experiences.

Chapter 2.

TEST REQUIREMENTS AND PURPOSE

Answering the question "What are we actually testing?" sets the stage for determining specific test objectives. This chapter presents a practical method that will help the manager or tester determine testing requirements. Again, let's assume that all DT&E and OT&E tests are complete and that we are testing for final acceptance. Simply stated, the purpose is to conduct one test or a series of tests and determine through system analysis the actual performance by comparing the results to specifications and usability requirements. There are several other standard tests that pertain to measuring system performance. These include the following.

2.1. Types of Tests.

- Complete system test. This is done when a thorough test that includes all interconnecting components and subcomponents is required. It allows the input stimulus to activate a system according to design and procedure, providing a snapshot of the system's performance and capability.

- Partial system tests. This type of test is used where full system connectivity is not required or feasible. The system is run just as it would be for a complete system test but with a single string configuration. It's the same concept as computing the reliability of a string of components in a system analysis for the weakest component link. The purpose of this test is to obtain an accurate assessment of system performance within a shorter period of time without comprising technical aspects.[1]

- Load or stress testing. In this type of testing, multiple subsystems are on-line simultaneously and the performance characteristics of one element are tested. This type of testing is necessary when determining system load capacity.[2]

- Regression testing. This type of testing is widely used when a system component is replaced or modified. More specifically, it measures and evaluates system performance before and after a change. This testing is very common in the reliability profession and is very effective in human-rated systems that involve redundancy.

2.2. Determining Test Requirements.

Deciding on testing requirements can be a difficult process. Each test varies due to system configuration, resources, and test objectives. Table 2.1 outlines the management considerations regarding test types.[3]

Table 2.1: Determining Test Type

- Identify critical testing issues
- System to be tested
- Evaluation criteria
- Data recording and evaluation requirements
- System specification
- Applicable system interfaces
- Test environment
- Required stimulus

Addressing each item in Table 2.1 will help management determine which type of test is needed. Managerial and technical expertise from internal and external sources can be useful during this critical step. Management and testers must work together to ensure that the most complete and properly planned system test is executed.

2.3. Application Example.

To meet user requirements for the RDS-120 tracking system, a complete system test that includes all ancillary equipment is most appropriate. That decision is based on the following responses to the criteria listed in Table 2.1.

- Identify critical testing issues. Can the RDS-120 tracking system acquire, track, and report a maximum of 20 aircraft 120 nautical miles from azimuth center?

- System to be tested. The RDS-120 tracking system, located at the Radar-Systems, Inc. test facility, will undergo a complete system test. The RDS-120 tracking system will be tested in an on-line operational environment.

- Evaluation criteria. Test objectives will reflect technical requirements and the system's technical specifications. Operators will provide a usability assessment.

- Data recording and evaluation requirements. Test data will be recorded at designated points along the radar chain (system) according to system capabilities (see Figure 3.1 on p. 20).

- System specifications. List the system specifications:
  System: RDS-120 tracking system
  Range: 120 nautical miles
  Load: 20 aircraft within simultaneous radar coverage
  Azimuth: 360 degrees of coverage (horizon to horizon)

- Applicable system interface. List all interfacing equipment and systems.
  Primary systems: radar receiver, intermediate processing equipment, central processing unit, and operational displays-Primary
  Ancillary Systems: operational display-Secondary

- Test environment. The RDS-120 tracking system will be tested on-line, tracking real aircraft penetrating the radar's 360 degree and 120 nautical mile coverage.

- Required stimulus. Actual aircraft tracks will provide radar detection returns to the radar receiver. This will cause normal system response.

2.4. References.

1. System Integration Office (SIO), Directorate of Test and Certification, SIO Standard 3100: *Test Capability Standard* (unclassified and released for publication on August 23, 1990), November 29, 1985.

2. Ibid.

3. System Integration Office (SIO), Directorate of Test and Certification, SIO Standard 3000: *Test and Demonstration Standard* (unclassified and released for publication on August 23, 1990), May 30, 1986.

Chapter 3.

TEST OBJECTIVES

To determine the test objectives, it's important to first understand the critical issues regarding the test. Critical issues address a system's essential capabilities, risk, or uncertainties that must be explored before its overall performance can be assessed. There are three basic types of critical issues: programmable, technical, and operational. Programmable issues, which mainly involve program schedules and resources, are addressed by the test developer. Technical issues usually involve the system's ability to meet key design specifications during the DT&E phase. Operational issues involve key system performance characteristics relating to effectiveness and suitability.

The critical operational test issues are sub-divided into objectives, which are in turn broken down into sub-objectives. These sub-objectives must be developed in the context of the overall purpose of the test. They must be planned to produce test results which, when properly completed and integrated, will support the objectives.[1]

3.1. Objectives.

The test objectives address the scope of the test and the critical issues, and ultimately determine the resources required. By examining each test, the manager can determine which issues are related and which can be placed in various interest groups such as reliability, inter-operability, and maintainability. Objectives should be mission-oriented and, when they're met, should provide information about the critical issues in the group.[2]

3.2. Sub-Objectives.

Test objective development begins with a comprehensive understanding of the purpose and intent of the test and the system to be tested. Developing the sub-objectives requires the combined skills of the operators, analysts, and engineers. The next step is to reconstruct the objectives by building sets of sub-objectives which, when assessed collectively, represent each objective. Sub-objectives must be identified and stated so they can be measured and translated into practical test and evaluation methods and procedures with operational, physical, and statistical meaning. Each sub-objective should be traceable as a direct contributor to one or more critical issues.[3]

Tables 3.1 and 3.2 will help management develop the objectives and sub-objectives respectively.[4]

Table 3.1: Determining and Evaluating Test Objectives

- Are critical issues adequately supported by test objectives?
- Are test objectives specific? Are they unique?
- Are test objectives consistent with the test purpose?
- Can the objective be satisfied through system testing?
- Are there too many test objectives?

Table 3.2: Determining and Evaluating Sub-Objectives

- Is the mission adequately addressed?
- Are the objectives consistent with existing or planned concepts and techniques?
- Do the objectives state the analysis requirements?
- Are reliability, maintainability, and availability properly considered in developing objectives?
- If software is apparent, is it adequately addressed?
- Do the objectives address areas of interface with other systems?
- Are environmental aspects properly factored into the objectives?
- Are test objectives developed for the environment?
- Are the objectives and sub-objectives realistic?
- Can any of the stated objectives be satisfied by other objectives?
- What is the proper stimulus?
- Is the software program designed to eliminate and control variables that will invalidate test results?
- Is the software proofed, validated, and controlled by revision levels?

Developing sound test objectives and sub-objectives is a critical step in the overall test and evaluation process. Maintaining the integrity of the objectives is essential. Management must ensure that testers properly execute the test to ensure the integrity of the test and that the objectives are achieved. When there are multiple objectives, the primary objective remains the driving force, in the order of sequencing the test events.

3.3. Application Example.

Developing and Applying Test Objectives for a Surveillance Radar Detection System (RDS-120 tracking system)

This example will focus on the application of five test objectives: detection accountability (DA), data error rate (DER), data receipt time (DRT), operational display accuracy (DA), and operator usability. The following example, shown in Figure 3.1, is generic and consists of the following individual components.[5]

- Radar receiver (RR). Receives input data from the radar antenna. These radar data are unprocessed returns or signals from acquired aircraft tracked and detected within the radar's area of coverage.

- Intermediate processing equipment (IPE). Acts as a buffer for radar return messages awaiting processing and reformatting by the central processing unit (CPU) into displayed operational data.

- Central processing unit (CPU). The mainframe computer system consisting of required system hardware, firmware, and software.

- Operational display (OD). The actual displays monitored by the operational user. The final portion of the radar chain that displays processed radar data (usable display) in real time to the operator. Primary displays are on-line continuously when the system is activated. Secondary displays present the same information but remain in a standby configuration until activated.

3.4. Discussion.

Test directors and managers must be knowledgeable of the system to be tested. This will ensure that adequate test objectives are developed and used in the testing. This example will assume 20 airborne aircraft penetrating radar coverage within a six-hour time period. The RDS-120 tracking system is capable of detecting all 20 aircraft according to design specifications.

Let's examine the five individual test objectives discussed and the impact of each on the system evaluation process.[6]

1. Data accountability (DA). The actual number of radar returns received by the radar compared to the actual number of displayed aircraft tracks on each operational display (includes primary and secondary). This objective is the accounting of data input to final processed and displayed information.

   Example: During the six-hour period, 20 aircraft penetrated the radar surveillance coverage as known detections. This first test objective evaluates and answers the question, "Did the radar scan, detect, acquire, and track the correct number of aircraft during the specified time period?"

2. Data error rate (DER). This test objective deals with the reliability of data generated and processed throughout the radar chain. This includes RR data; input, processing, and output of CPU data; and displayed data.

   Example: Performance is measured against the system's ability to differentiate and correctly process raw data returns into usable displayed events. If not, the software quality or hardware components might become questionable or require modification and/or upgrade to meet system demands.

3. Data receipt time (DRT). The total elapsed time from detection to presentation of information on the operational displays, including CPU processing time.

   Example: As the number of aircraft within coverage increases over a given time period, CPU processing time will increase. Measure the time lapse between each system component (time stamp), for instance the time lapse from RR to arrival at the IPE. A sample formula for computation of total receipt time is:

   Time lapses between: RR and IPE + IPE and CPU + CPU and OD * = Data Receipt Time (DRT)
   *includes display update time

4. Display accuracy (DA). This test objective involves display verification of the actual number of aircraft detected. It takes into account clarity, completeness, and accuracy to include operator-usable considerations for primary and secondary displays. This particular test objective answers the questions, "Is the RDS-120 tracking system presenting all required data accurately to the operator?" and "Are both primary and secondary displays updating in a timely fashion?"

5. Operator usability (OU). Measured with the evaluation of four sub-objectives that are necessary in determining operator suitability. Sub-objectives are best presented in question formats as follows:

- Is the operator-machine interface acceptable? This is an evaluation of the system's user-friendly abilities. This sub-objective should cover operator adaptability and training effectiveness.

- Can the system handle multi-tasking? System specifications should allow for the simultaneous tracking and process of multiple aircraft detections. This sub-objective is necessary when determining the system load capacity as described in Chapter 2.

- Is a redundant system capability available and usable? Most radar surveillance systems have redundancy that activates automatically or manually when the system recognizes or experiences an internal, hardware, software, or firmware failure. The effectiveness of operations, length of switching time, and system normalization should be the main criteria for measurement. In addition, determining and identifing the "single point failure" within the system should also be addressed. This is regarded as a critical failure involving a single component interconnected with all remaining system equipment.

- Are the secondary displays similar to primary displays? Does each display provide similar data and update capability? System redundancy is an important sub-objective.

In conclusion, when testing the radar detection system, managers and testers must have a complete understanding of system operations to develop accurate test objectives and sub-objectives. In this example we examined five objectives and four sub-objectives (providing evaluation for the fifth test objective) and applications to operational effectiveness and system performance. To ensure an accurate assessment of system integrity, test objectives must also factor the interoperability of each radar component.

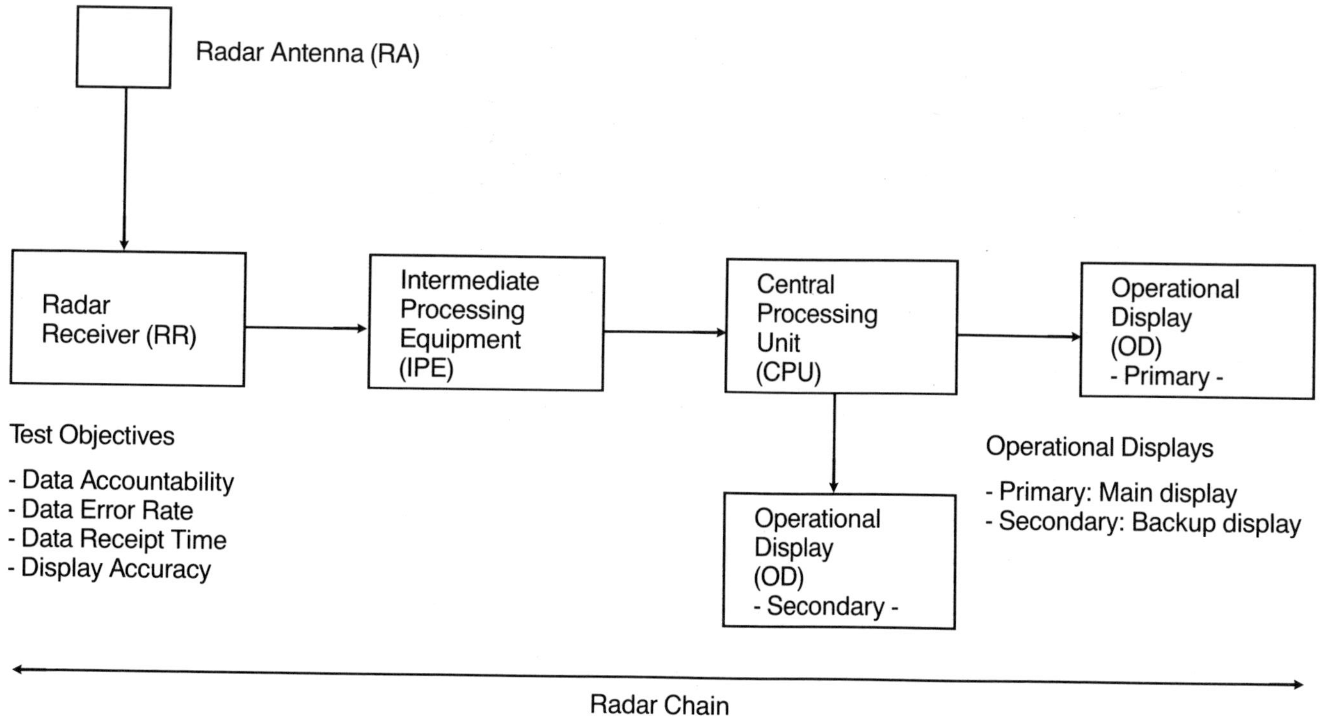

Figure 3.1: Surveillance Radar Detection System. Hypothetical diagram indicating specific portions of the surveillance radar detection system. This diagram is unclassified and based on available SIO test documentation. Released for publication on August 23, 1990.

3.5. References.

1. Department of the Air Force, Air Force Regulation 55-43, *Management of Operational Test and Evaluation* (unclassified and released for publication under public domain), Air Force Operational Test and Evaluation Center, June 28, 1985.

2. Ibid.

3. Ibid.

4. Author's testing experience and knowledge while assigned to the System Integration Office (SIO). This experience extends to the author's first-hand involvement with the designing, planning, coordinating, and managing of annual ballistic missile and atmospheric and space tactical warning and attack assessment (TW/AA) system end-to-end (from face of sensors to all command centers and forward world wide users) high load performance tests. The objective of these tests was to assess the overall technical integrity of the entire systems utilizing "test" or "non-real" events in a controlled test environment.

5. Ibid.

6. System Integration Office (SIO), Director of Test and Certification, *SIO Standard 3100: Test Capability Standard* (unclassified and released for publication August 23, 1990), November 29, 1985.

Chapter 4.

TEST CAPABILITIES

The actual test capabilities of the system must be considered in the initial test planning and development phase. The main concern regarding test capabilities can be summarized as: "How can the tester activate the system in a test configuration to simulate a real environment?" Naturally, the more realistic the system simulation, the more effectively the test results can be measured against real-world operational requirements.

Table 4.1: Determining Test Capabilities
• Test data injection points • Processing test data • Displaying test data • Test data dissemination • Interrupting test data processing • Synchronization, coordination • Test preparation time • Simulation, software • Message recording and reduction

Table 4.1 shows the key factors to be addressed when determining test capabilities. When this step is completed, the design of the test will reflect test concept, and the objectives of the test will be met.[1]

4.1. Application Example.

Employment of Test Capabilities for the Surveillance Radar Detection System (RDS-120 tracking system)

This example will discuss the application and testing limitations of specific test capabilities available for use against the hypothetical surveillance radar detection system. The study is representative of many systems and will cover stimulus injection locations and resulting outputs at each radar system component.

Figure 4.1 shows the configuration of the surveillance radar detection system, with the stimulus injection locations along the radar chain.

4.2. Discussion.

The test capabilities of this system allow for off-line system testing during scheduled and unscheduled maintenance and during system testing time. After each maintenance activity, a complete or partial system test should be conducted to ensure proper operation and system integrity. By factoring in the type of system change against specific test capabilities, the tester must be familiar with the required stimulus, test injection system, appropriate injection locations, and expected system output. This is necessary to effectively test against established test objectives and system specifications. Test capabilities are applied in Figure 4.1.

Data injection points correspond to the following system locations indicated in Figure 4.1.[2]

- Location A. The RR injection point is located after the antenna system and inputs directly into the RR. System test capabilities should include equipment that can generate or simulate raw radar return data as close to the RA. Proper test data injections at this location can give the tester an entire off-line and on-line system assessment excluding radar antenna inputs.

- Location B. The IPE injection comes after the RR output and before the IPE input. IPE data injection assumes proper operation of the RR equipment. The main function of this IPE is to hold (buffer) RR data, then sort and transmit these data directly into the CPU. The IPE buffer software releases RR data based on CPU data processing availability. From this location the tester can assess the IPE buffer capacity to include CPU and OD loading and display update capability, respectively. The test data injection system must have the capacity to duplicate and output RR data.

- Location C. The CPU injection comes after the IPE output and goes directly into the CPU input. CPU data injection assumes proper IPE data format output. Different CPU input patterns can provide a wide range of loading or stress testing of the CPU software. Since the CPU is where most software and firmware changes occur, this capability has a high test value to the software developer by reducing off-line system test time. From this location effective testing of the CPU and OD is possible.

- Location D. OD input considers only processed CPU data formats. Data are input directly into the OD and resemble processed CPU output information.

In conclusion, to enhance testing flexibility, the total dependency on real-world stimulus should be avoided. System test capabilities are necessary to facilitate accurate trouble-shooting and realistic off-line system testing. Test equipment should allow for proper data input peculiar to each system component. The lack of system test capabilities can inhibit an accurate system performance evaluation. When analyzing the suitability of the RDS-120 tracking system, on-line and off-line test capabilities must satisfy user requirements. Managers and testers must carefully consider this capability, since it directly affects test time and configuration, system availability, maintenance costs, and scheduled maintenance times.

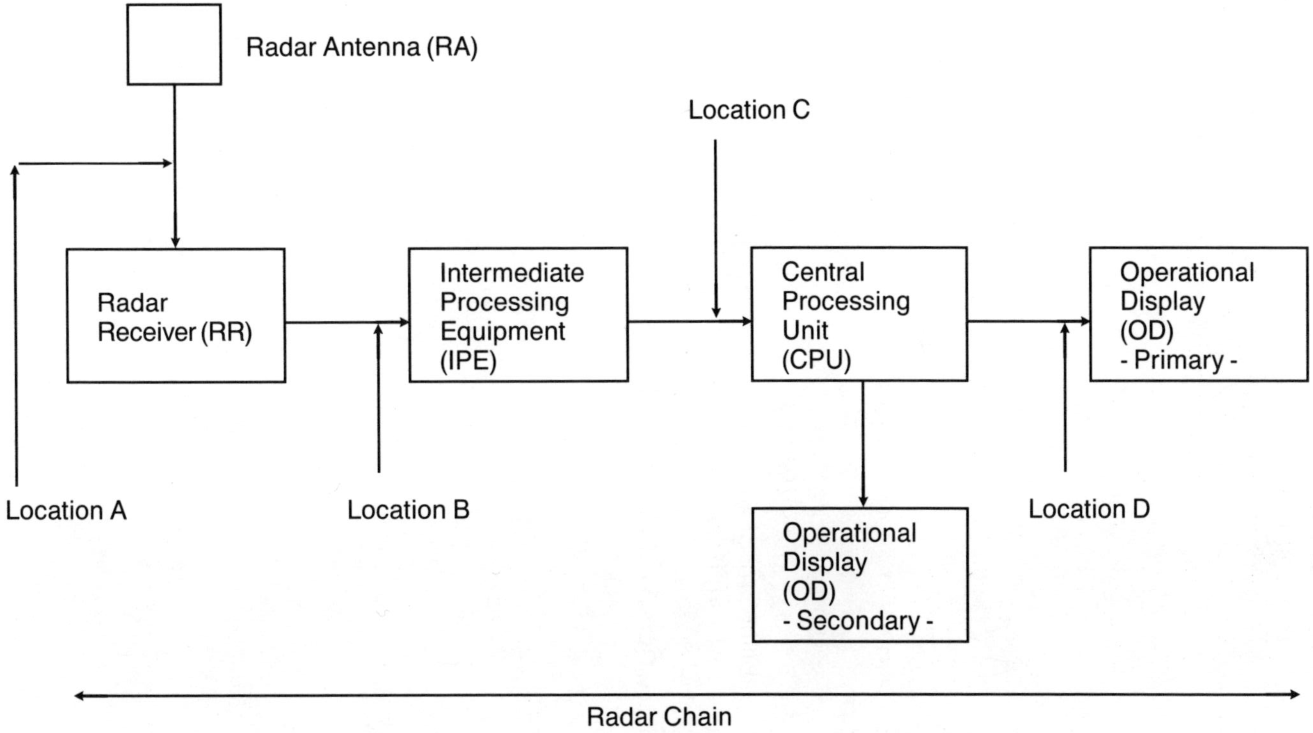

Figure 4.1: Test Capabilities for the Surveillance Radar Detection System. This diagram indicates practical examples of various data injection locations along the surveillance radar detection system chain. This diagram is based on the author's testing experience and SIO testing documentation. Approved for publication on August 23, 1990.

4.3. References.

1. System Integration Office (SIO), Directorate of Test and Certification, *SIO Standard 3100: Test Capability Standard* (unclassified and released for publication August 23, 1990), November 29, 1985.

2. Ibid.

Chapter 5.

DESIGNING A TEST

This chapter will provide a practical method of designing a test based on the requirements and the purpose. Test design is nothing more than developing a plan of attack to conduct an experiment, gather essential data, analyze test data, and derive conclusions with a minimum expenditure of test resources.

The actual test design characterizes the purpose of the test, the required steps for successfully performing the test, and what can be learned from the test. Thus, a carefully planned and designed test has a greater chance of satisfying all test conditions and areas of concern. If it's not properly planned and designed, the test will be of limited use. In short, the test design represents how good the test can be. A good test design cannot guarantee an effective test, but a poor test design guarantees an invalid and ineffective test, no matter how carefully it is executed. Table 5.1 outlines basic managerial considerations regarding a practical test design.[1]

Table 5.1: Test Design Considerations

- Simplicity
- Account for all sources of variability
- Avoid controlled environments for operational tests
- Determine proper operational environment
- Factor all variables that might affect outcome
- Define the dependent variables
- Identify independent variables
- Determine test control factors, both single and multiple
- Speculate on test results for effectiveness determination

A properly planned and conducted test design provides the maximum amount of useful information from available resources and helps management isolate the effects of dependent variables or measures of effectiveness (MOEs) identified in the test plan. An effective design reduces the margin of error by combining test conditions, allowing for more accurate and reliable information about the relationships among important variables.[2]

Once all dependent variables are identified, all independent variables that affect the outcome of the test must be accounted for. Primary control factors are of special interest to the test manager or test designer because of their effect on dependent variables. Background control factors are independent variables that cannot be controlled with regard to the operational environment.

Single primary factor designs are distinguished by only one primary factor within the test. For example, a randomization in test design will even-handedly distribute sequence-correlated events. Other types of single primary factor designs exist to account for different combinations of background factors and randomization.[3]

Multiple primary factors have more than one primary factor. When the number of viable factors has been reduced, a closer inspection of their relationships becomes more feasible and economical. Partial factorial designs go even further to economize on test points and resources by eliminating higher-order interactions. Thus, we can spend a few resources up-front in a well-prepared factorial design to isolate those variables that most affect the variable of interest and then concentrate our resources in those areas.

Once the tester has determined purpose, requirements, design, and variables and included each item in the test plan and procedures (discussed in Chapters 6 and 7), enough information should be available to hypothesize expected results. This practice will provide effective feedback when constrasting actual test results to system specifications or known operator requirements.

5.1. Application Example.

Since the owner and manager of Rocky Mountain Airport wants to design and conduct the test, basic managerial considerations apply. An appropriate test for this system can be effectively designed according to the requirements shown in Table 5, as follows.

- **Simplicity of variability.** Number of aircraft penetrating the radar's coverage. This is an on-line system test.

- **Avoid controlled environments.** All testing is performed in an operational environment.

- **Determine proper operational environment.** On-line system located at Radar-Systems Inc. test facility. Access to operational airport configuration.

- **Factor all variables that might affect outcome.** General examples: system failure, weather, availability of aircraft, learning, and fatigue.

- **Define the dependent variables.** These are measures of test outcome. More specifically, the test objectives and sub-objectives.

- **Identify independent variables.** Variables that might affect the dependent variables. These variables can be controlled or uncontrolled.

- **Determine test control factors.** The RDS-120 tracking system will be operational. Although test capabilities allow for system loading, real-world aircraft tracks are primary test stimuli.

- **Speculate on test results for effectiveness determination.** Detect, track, display, and record all aircraft tracks.

For additional information on test design, consult the sample Test Preparation Checklist in Appendix 1. This managerial tool provides the basic ingredients for identifying test design requirements; developing and reviewing the test plan and procedures, test execution actions, Quick-Look report actions, data collection and analysis actions; test reporting; and lessons learned considerations. Managers who are testing several systems can benefit by adding specific requirements to this example, which will result in a more comprehensive checklist for each system.

5.2. References.

1. Directorate of Analysis, Air Force Operational Test and Evaluation Center, *Technical Paper 3.0, Test Design* (unclassified and released for publication on November 7, 1990), Kirtland Air Force Base, New Mexico, May 1986.

2. Ibid.

3. Ibid.

Chapter 6.

DEVELOPING THE TEST PLAN

Test plans and procedures can take on various formats. This chapter and Chapter 7 will highlight the important elements of a test plan and test procedures. In general, a test plan should serve as a managerial control document for test preparation and execution, as well as the reporting of test results. The minimum contents of the test plan are the test purpose, schedule, participants, configuration, direction and location, procedures, execution, aborts and restarts, operational effects, data collection, test and anomaly reports, and a glossary.[1] A good test plan provides an executive overview from test purpose all the way through to posttest analysis and reporting.

1. Test Purpose. This is a statement of the actual test and the reason for conducting it. The test purpose must be clearly stated.

2. Test Schedule. This should indicate the time, place, duration, and responsible agency for conducting the performance test. The test plan serves as the official test notification to all participants. See the ''Test Participation Notification'' in Appendix 2.

3. Test Participants. Responsible agencies or individuals are listed to include specific functional areas. List everyone from administrative support to participating agencies. This will assist management with test coordination, task assignments, materials, and reports.

4. Test Configuration(s). This diagrams all operational interfaces and test equipment used in the design of the test. This includes, for example, how test stimuli are injected into the system, the data paths that are activated, and the expected system output. It also includes specific locations throughout the system where test equipment is used to provide required readings for performance analysis.

5. Test Direction and Location. Indicate the test director and test location for all participants.

6. Test Procedures. This is a checklist of specific actions used to conduct and execute the test. Most test procedures are issued under separate cover from the test plan (the test plan serves as an executive overview of test requirements). Test procedures will provide a specific step-by-step method to be used for test configuration, scenario execution, and reconfiguration to normal operations. All test participants indicated in the test plan must have a copy of the test procedures. A trial run of the test procedure to ensure its effectiveness is desirable.

7. Test Execution. Specifically state the stimulus and method of test execution. The detailed test execution requirements are included in the test procedures. How will the test start?

8. Aborts and Restarts. Indicate the circumstances and individual(s) responsible for stopping and restarting the testing sequence. This section should also indicate the test degradation limits that result in an abort or restart.

9. Operational Effects. The specific system limitations must be defined so that alternative procedures and contingency planning can be made available to maintain partial or complete operational capability. This pertains to primary and ancillary systems so that no mission capability is adversely affected by these testing procedures.

10. Data Collection Requirements. The data collection requirements must be described in detail for each primary and ancillary system component involved with the testing. The test manager must clearly indicate the exact test period for collecting data, the media used, special markings and format, etc. Data collection requirements pertain to all test executions regardless of duration and the number of executions. A sample "Data Collection Checklist" is shown in Appendix 3.

11. Test and Problem Reports. All participants are required to document observed test data and complete a test report. The format for this report should reflect the expected test output. Each test participant will complete an anomaly report for each occurrence or condition affecting test execution or data collection, and will record their observations of the test circumstances. See Appendix 4 for a sample "Test Report Checklist" and Appendix 6 for a sample "Problem Report."

12. Glossary. The following list of terms and abbreviations provides a sample reference for managers. This particular list is not all inclusive and contains standard definitions. When a test plan is being developed, this section of the test plan is critical because it standardizes the terms.[2]

Administrative support: Secretaries, clerks, reproduction facilities, typist, mailroom, clerks, etc.

Analysis: The detailed examination and application of disciplined techniques in order to understand its nature and essential features.

Ancillary systems: Subsystems not under direct evaluation but affected by system output and the performance of primary systems. Same term as secondary, interconnecting, subsystem, and peripheral equipment.

Availability: A measure of the degree to which an item is operable at the start of a mission when the mission is called for at an unknown (random) time.

Background control factors: Other independent variables that cannot be controlled in the operational world and will not affect the system if employed. There are three ways to control the effects of background factors on the dependent variables: hold constant, randomize, and blocking.

Blocking: A block is a portion of the experiment that is expected to be more homogeneous than the aggregate. For example, if we wanted to test the wear characteristic of four brands of tires, we would place a different brand on four separate vehicles.

Combined testing: Testing conducted by the development and operational testers to conserve resources, compress schedules, and reduce costs. This is not a recommended method of testing, since DT&E, OT&E, and acceptance testing should be separated.

Compatibility: The capability of two or more operational items or systems to exist or function as elements of a larger operational system or operational environment without mutual interference.

Controlled test environments: The system test configuration depicting or simulating actual operational conditions for software, firmware, or hardware components.

Critical issues: Those aspects of a system's capability, either operational, technical, or other, that must be questioned before its overall worth can be estimated and that are of primary importance in deciding whether the system will be advanced to the next acquisition phase.

Decision authority: The agency or individual responsible for determining user requirements and acceptable performance.

Dependent variables: Specific factors that can be controlled by the tester to stage a proper test environment. For example, scenario to be used, test date, or participants.

Development testing and evaluation (DT&E): That testing and evaluation of preproduction articles used to measure progress; verify the accomplishment of development objectives; determine if theories, techniques, and material are practicable; and if systems or items under development are technically sound, reliable, safe, and satisfy specifications.

Downtime: For a specified period of time, the total time a system is not operational.

Evaluation: The review and analysis of qualitative or quantitative data obtained from design, review, hardware inspection, testing, or operational use of equipment.

Evaluation criteria: Standards used to judge the achievement of required operational effectiveness or suitability characteristics or the resolution of technical or operational issues.

Facilities: This includes system-unique structures required for system support.

Firmware: Computer instructions and/or computer data that resides on a hardware device as read-only software.

Goals: Levels of performance (established by the user) above those required which, if achieved, will provide additional operational capability.

Hardware: Computer hardware relating to configuration, connectivity, design, or physical location of any component. The physical system(s) used to execute computer software routines.

Hold constants: Keeps variability effects from influencing dependent variables. For example, picking aircraft tracks representative of the range of real world tracks.

Human factors: Those factors that contribute to the optimization of a system by integrating the human performance necessary to operate, maintain, support, and control the system in its intended operational environment.

Independent variables: Specific factors that cannot be controlled by the tester to stage a proper test environment. Examples are weather, mission requirements, workschedule, etc.

Initial operational test and evaluation (IOT&E): The first phase of operational test and evaluation conducted on preproduction items, prototypes, or pilot production items and normally completed prior to the first major production decision. It is conducted to provide a valid estimate of a system's operational effectiveness and operational suitability before the first major production decision.

Injection points or locations: Software provided to inject simulated events into a system. The injection point for the simulated returns shall be as close as practical to the injection point for actual event processing.

Interoperability: The ability of systems and human factors to provide services to and accept services from other systems or units and to use those services to operate effectively together.

Measures of effectiveness (MOE): A quantitative or qualitative measure of a system's performance or characteristic that indicates the degree to which it performs a task or meets an objective under specified conditions.

Primary control factors: Factors that have an effect on the dependent variable. These factors can be controlled by setting them at various levels during the experiment.

Primary systems: Specific systems components directly under performance evaluation.

Randomize: This technique minimizes variation by producing samples representative of the population of interest. Randomization is often easier to do in theory than in practice because of real-world constraints.

Realistic test environment: The conditions under which the system is expected to be operated and maintained.

Reconfiguration: System posing an on-line/off-line test capability might require normalization or switching from the off-line to on-line system and vice versa.

Reliability: The probability that a system, subsystem, or equipment will perform a required function under specified conditions without a failure for a specified period of time.

Required operational characteristics: System parameters that are primary indicators of the system's ability to be used to perform the required mission function and to be supported.

Required technical characteristic: System parameters selected as primary indicators of the achievement of technical goals and specifications.

Requirements: Minimum levels of performance or capability at planned system maturity describing a mature system value that will allow satisfactory accomplishment of the system's operational mission.

Responsible agencies: Organizations that are directed via test requirements to participate in testing.

Scenario execution: The orderly sequence of events providing the proper test stimulus for the system test. A scenario can consist of pre-recorded magnetic media, or manual, mechanical, or visual stimulus. A scenario usually is a programmed sequence of inputs stimulating a specific system response. Similar to the term "testing sequence."

Software: The resident instructions that cause a computer to execute an operation.

Software maintainability: Includes those characteristics of software that affect the ability of the software programmer to correct errors, enhance system capabilities through software changes, and modify the software to be compatible with hardware changes.

Task assignments: Specific work schedule for each test participant. This involves work assignments for pretest, execution, and posttest requirements.

Test degradation: The designated point or subjective determination by the tester about test configuration or execution that affects test requirements and purpose.

Test director: The person assigned to conduct a test according to an operational test and evaluation (OT&E) plan. This individual exercises overall responsibility for achieving OT&E plan objectives.

Test manager: The person designated as the focal point for advance planning, executing, and reporting on an OT&E test program.

Verification: The process by which a technical order is tested and proven.

Verify: An effort to conclusively establish the correspondence, truth, accuracy, or validity of a contention.

6.1. Application Example.

The following is a "Test Plan Example" for the RDS-120 tracking system based on the outline provided in this chapter.

1. Test Purpose. Test the RDS-120 radar system to demonstrate that the operational and system technical specification requirements are suitable and attainable. To verify the usability of this system, an on-line operational environment is required.

2. Test Schedule. The primary date for the start of operational testing is January 2 1991. The secondary date is January 3 1991. Testing will begin at 8 a.m. at the Radar-Systems Inc. test facility.

3. Test Participants. Make a list of all employees, testers, contractors, and administrative staff associated with and participating in this test. (See the sample "Test Participant Notification" in Appendix 2.)

4. Test Configuration. See Figure 3.1. Include a basic system level diagram if necessary to visually present the system under test and evaluation.

5. Test Direction and Location. The test director is Michael J. Muolo; location is the Radar-Systems Inc. test facility.

6. Test Procedures. See Appendix 5 for sample procedures.

7. Test Execution. The RDS-120 tracking radar will be configured in an on-line operational mode. Real-world aircraft penetrating the surveillance radar coverage will be tracked from in-bound and landing status to takeoff status from the operational runway. Rocky Mountain Airport operators will operate the RDS-120 equipment under the direct operational supervision of Radar-System Inc. technicians. Test data will be recorded at four specific system locations as indicated in Figure 4.1.

   This test is scheduled for six hours, allowing for 10 scheduled aircraft arrivals and five departures. Test data will be reduced and analyzed to determine operational performance based on established test objectives. See Appendix 3, "Data Collection Checklist."

   Once test data are analyzed they will be compared to system specifications for performance and operational suitability requirements.

8. Aborts and Restarts. This test will be rescheduled due to equipment malfunctions. The test will not be postponed in the event of adverse weather conditions.

9. Operational Effects. The secondary operational display will be activated should the primary operational display fail for whatever reason. A system restart will be required should a complete or partial system failure occur. Test data will not be recorded during this restart. (All test data not stored on disk will be lost when the restart is initiated.) Aircraft tracks will require requisition should complete or partial system failure occur. After restart is complete, ensure that the latest data are recalled before tracking resumes. This will provide the radar operator with the last known aircraft position.

10. Data Collection Requirements. Outputs must be recorded in accordance with Appendix 3. Data must be recorded in a readable format, and all real-time test observations must be documented.

11. Test and Problem Reports. During the test, record all system anomalies for incorporation into the quick-look report, test analysis, and final test report. See Appendix 6 for a sample "Problem Report."

12. Glossary. List terms and abbreviations.

6.2. References.

1. System Integration Office (SIO), *Combined Ballistic Missile and Atmospheric TW/AA System End-to-End Performance Test Procedures* (unclassified and released for publication on November 23, 1990). Actual test procedures pertaining to the Ballistic Missile and Atmospheric Warning System configuration was not referenced. Information pertaining to procedure format was referenced. August 12, 1988.

2. Department of the Air Force, *Air Force Regulation 55-43, Management of Operational Test and Evaluation,* (unclassified and released for publication under public domain), Air Force Operation Test and Evaluation Center, June 28, 1985.

Chapter 7.

DEVELOPING TEST PROCEDURES

A test procedure should be unique and give the tester and other participants a complete step-by-step method of establishing the test environment. Developing and distributing the test procedures is the final step in the pretest requirements prior to test execution (see Figure 7.1). Test procedures should be clear and concise, allowing flexibility for the tester to compensate for unexpected occurrences.

The test procedure must detail the specific action required of each participant, and this information must be made available in advance. Since the test procedures are critical to a successful test, the participants should review them in advance. The test procedures must reflect every detail from test coordination all the way through to reporting the test results. The test manager must control the testing by preserving test integrity and approving any recommended changes to the procedures during the review process.

Test coordination packages are specific instructions for all test participants that include checklists for test data collection and test and problem report forms. Providing these support packages standardizes data collection requirements, simplifying the work of the person conducting the test.

The managerial considerations shown in Table 7.1 are necessary when developing test procedures.[1]

Table 7.1: Developing Test Procedures

- Test type
- Required system operators
- Operator responsibilities
- Operator and machine tasks
- System communications
- Maintenance philosophy

This chart is not all-inclusive, but it provides the focused direction that's necessary for developing test procedures. Test procedures should be developed by the system experts. The minimum sections necessary to develop effective test procedures are as follows:[2]

- Test Procedure Purpose. Includes test purpose, participants, contents of test support forms, and points of contact.

- Test Procedures Organization. Contains information of interest for all participants. Each participant is identified and assigned to respective sections within the procedures.

- General Information. Includes method of test execution, participants, data collection requirements, communications networks, scenario description, and termination procedures.

Test procedures and data collection tools should be reviewed and, if possible, undergo a "trial run" to ensure their validity.

7.1. Application Example.

See Appendix 5, "Test Procedures Example for RDS-120 Tracking Radar."

7.2. References.

1. Author's testing experience and knowledge while assigned to the System Integration Office (SIO). This experience extends to the author's first-hand involvement with the designing, planning, coordinating, and managing of annual ballistic missile and atmospheric and space tactical warning and attack assessment (TW/AA) system end-to-end (from face of sensors to all command centers and forward world wide users) high load performance tests. The objective of these tests was to assess the overall technical integrity of the entire systems utilizing "test" or "non-real" events in a controlled test environment.

2. System Integration Office (SIO), *Ballistic Missile Tactical Warning and Attack Assessment End-to-End System Performance Test Plan for (High Load Scenario Run No. 5)*, (unclassified and released for publication November 23, 1990), August 12, 1988.

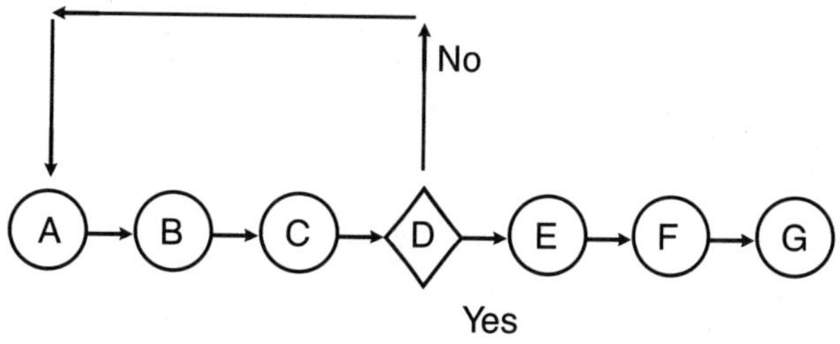

A. Test Requirements and Purpose
B. Test Objectives
C. Test Capabilities
D. Is the test feasible?
E. Designing A Test
F. Developing A Test Plan
G. Developing A Test Procedure

Figure 7.1: Final Acceptance Testing: Pretest Requirements. This diagram is representative of the author's method of coordination and standard testing procedures. Approved for publication on November 23, 1990.

Chapter 8.

TEST EXECUTION

This chapter discusses the requirements for test execution and assumes that all test preparations are accomplished. On the day of the test the test manager will review all test procedures to ensure that the test requirements, system configuration, participants, and test control functions are ready. Before the test begins the test manager should also hold a staff meeting to discuss, coordinate, and review all concerns about the test. This includes the test requirements and purpose, test objectives, test capabilities, test plan, test procedures, and the manner of test execution. This preparatory meeting will provide excellent coordination for the review of data collection and quick-look report requirements. One efficient method is to use a test execution checklist to ensure that all pretesting requirements are satisfied. It is not unusual, during the test configuration, for equipment problems to arise, resulting in a deviation from the planned test configuration. At the staff meeting, the test manager should discuss configuration and execution alternatives to decide on an acceptability range before declaring a test invalid or incapable of being conducted.

The following questions will help the participants review pretest functions and stimulate ideas for alternative action when required. The checklist is grouped into five major sections: confirmation of test procedures, 30-day notification to participants, seven-day notification to participants, final 24-hour contact of participants, and test execution day.

Confirmation of Test Procedures

1. Have all participants received test plans and procedures?
2. Did participants make changes to the test plan and procedures?
3. Were changes made to the test plan and procedures, and retraining conducted?
4. Did all participants receive the data collection packages?
5. Are all participants aware of the date, time, and location?
6. Do all participants know how to contact the test manager?
7. Are all test participants capable of executing test procedures?

30-Day Notification

1. Have test participants responded to the 30-day test notification?
2. Have all participants indicated points of contact, phone numbers, mailing addresses, and position information?

Seven-Day Notification

1. Have test participants responded to the seven-day test notification?
2. Are there changes in participant status since the 30-day notification message?

3. If equipment downtime is appropriate for this test, has it been scheduled?
4. Are alternative operations available? If not, are personnel aware of reduced or limited operating capability?
5. Are the test director and managers prepared for this test?

Final 24-Hour Confirmation

1. Have final details been satisfied?
2. What is the equipment status for the test?
3. Have procedural changes been instituted?
4. Are data recording available and ready at each output location?
5. Has the test manager contacted all participants?
6. Are the test director and manager ready to conduct the test?

Test Execution

1. What is the equipment status?
2. What is the final test configuration?
3. Have the test objectives and critical issues been reviewed?
4. Have the test procedures been reviewed?
5. Do all participants know the test start and termination times?
6. Are there any questions on procedures or data recording requirements?
7. Are participants aware of how to make and distribute test data?
8. Is everyone aware of test abort and restart procedures?
9. Do all participants know about the posttest briefing immediately following the test?
10. Are participants aware of the need to submit a quick-look report to the test manager within the established time limit after test termination?

Remember, a well thought-out test execution will almost guarantee a successful test. If possible, a testing conference between all test participants should be conducted in order to maintain good test command and control. Maintaining telephone communication among test conferees helps in coordinating real-time test results while the test is being conducted. Conferencing capability is extremely useful for substantiating the quick-look report.

Refer to Appendix 4 for a sample Test Report Checklist and to Appendix 6 for a sample Problem Report Format, both applicable to this phase of the process.

8.1. References.

1. Author's personal testing experience and knowledge while assigned to the System Integration Office (SIO). This involves test executions that pertain to ballistic missile, atmospheric, and space tactical warning and attack assessment (TW/AA) system end-to-end (face of sensor to command centers and forward users) technical performance tests. Since these performance tests involve up to 125 world wide participants, notification of test execution is critical to successful test execution. In addition, "test" or "non-real" test environments must be considered with the test execution planning. This list represents a summary of the author's preferred action items.

Chapter 9.

QUICK-LOOK TEST EVALUATION

When a large system test is at stake the amount of data collected might exceed hundreds of thousands of individual automated listings, or individual data components, that require evaluation. Sometimes an automated analysis capability is available, resulting in a shorter turnaround time for analysis. The quick-look test report is a useful tool that provides an initial test assessment to management and participants. Some system tests might involve distant participants resulting in a waiting period for test output data. As a minimum, quick-look reports should contain the sections shown in Table 9.1.[1]

Table 9.1: Quick-Look Report Contents

- Start Time
- Test objective(s)
- General test description
- Problems observed during test execution
- Initial test results
- Termination time
- Objective summary

The quick-look report should have the following capabilities:[2]

- It must be possible to determine whether the system is functioning normally or whether a fault exists in near real-time so that excessive time is not spent collecting data that represent a malfunctioning system.

- It must be possible to determine whether instrumentation is operating properly in near real-time so that large numbers of measurements are not taken with instruments that are not working properly or are improperly calibrated.

- It must be possible to determine in near real-time whether a scenario and exercise simulation have been conducted according to plan so that collected data represent what was desired.

- Sufficient data on test results must be available in near real-time to permit decisions as to whether the test can be ended because the system being tested is clearly superior or inferior to the comparison system.

An important assignment for the tester is to continually observe the conduct of each test with realism in mind, and to quickly make any minor changes that will prevent a sense of artificiality. It is not possible to identify specific factors to watch for in this regard. However, the test director who is experienced in actual systems operations will immediately note anything that is out of place. It is that person's responsibility to take immediate corrective action or report initial results via quick-look reporting. Refer to Appendix 7 for a sample quick-look test report format.

9.1. References.

1. Quick-look report format designed by the author.

2. Roger T. Stevens, *Operational Test & Evaluation: A System Engineering Process* (Malabar, FL: Robert E. Krieger Publishing Company, 1986).

Chapter 10.

DATA COLLECTION REQUIREMENTS

Data collection is a critical component of the testing process. If the proper output data are not collected, the entire system test can be jeopardized. Test data can be produced in various media such as magnetic tape, hard copy printout, videotape recording of specific displays, the sequence activation of alarms or console indications, a specific system reaction or response, or even hand-written notes. The speed of the data analysis ranges from manual to automated, depending on the media and resources available. Whatever the case, data analysis involves the comparison of actual test results to established or system design specifications. During this portion of the testing process conclusions are drawn pertaining to the test objectives. Assumptions or "best guesses" produce unacceptable testing conclusions and must be discouraged.[1]

Remember that in the test plan data collection requirements will have identified each data item produced or calculated from some combination of test data and analysis tools. Each test objective inherently defines the required measurement or calculation to be made. In addition, how these measurements and calculations are combined will provide answers to the critical issues.

Data collection should satisfy the following objectives:[2]

1. Provide facts for drawing conclusions and making recommendations to system users.
2. Provide factual data for publication in the final test report in support of conclusions and recommendations.
3. Provide detailed factual data that can be referenced by anyone who wishes to question the conclusions, recommendations, or conduct the test.

Optimum data collection and processing methods are those which make it easiest to use the recorded data to accomplish these three purposes. Test data usually occur in one of two forms, either recorded or calculated manually by test observers, or recorded automatically by test instruments in the form of tapes, punched cards, or computer printouts. Tapes and punched cards will most likely require some type of data reduction in order to become readable.[3]

10.1. Manual Data Collection.

This is a simple process dealing with recording actual test measurements in real-time and determining the required conclusions. To efficiently record and process manual test data, a form designed specifically for the individual test should be developed, reflecting test objectives.[4]

10.2. Automated Data Collection.

This procedure is slightly more complex but more efficient when analyzing test data. In most cases the necessary calculations are made and provided to the test manager in real-time. Readability of refined test data is also critical to a successful test.

Manual data collection and process can provide a quick-look capability if carefully planned in conjunction with simple system performance tests. As new complex systems enter the design phase, program managers need to ask users and system operators what are the necessary automated data collection and reduction requirements.[5]

Refer to Appendix 3 for a sample Data Collection Checklist and Questionnaire.

10.3. References.

1. Roger T. Stevens, *Operational Test & Evaluation: A System Engineering Process* (Malabar, FL: Robert E. Krieger Publishing Company, 1986).

2. Ibid.

3. Ibid.

4. Ibid.

5. Ibid.

Chapter 11.

ANALYSIS OF TEST DATA

Depending on the type of test involved, the analysis effort must reflect the test objectives. Data recording efforts must also model the same objectives to ensure that the appropriate test data are collected. Statistical computations involving data sampling are more appropriate for tight schedules and quick turnaround of test results. Statistical sampling should be performed only when absolutely necessary or when feasible. The bottom line is that criticality is a factor of purpose and quality of data. This chapter will discuss performance analysis, determination of discrepancies, and reporting discrepancies. Management must decide on the appropriate analysis tools to employ for specific analyses.[1]

11.1. Performance Analysis.

This analysis technique must be applied to the test purpose and objectives. Consider the following:[2]

1. How large a shift do we wish to detect?
2. How large a sample, if any?
3. How critical the mission?
4. What is the margin for error?

11.2. Determine Discrepancies.

After analysis, compare the test data to test objective fulfillment. Differences that exceed the established error specification range can be corrected and considered discrepancies. Differences that cannot be corrected should be identified as system limitations and identified in future test plans.

11.3. Report Discrepancies.

If required, a formal configuration and control system should be established to identify, document, and direct the correction of system discrepancies. System discrepancies must be documented and the corrections to be made must be prioritized.

11.4. References.

1. Roger T. Stevens, *Operational Test & Evaluation: A Systems Engineering Process* (Malabar, FL: Robert E. Krieger Publishing Company, 1986).

2. Ibid.

Chapter 12.

THE TEST REPORT

It is essential that all testing be completely documented and assembled into a test report. The purpose of the test report is twofold:

1. To provide performance conclusions and recommendations to be used by top-level decision makers.
2. To provide a detailed description of the system, test methods, and test results. It also permits people who are unfamiliar with the system to reason out the conclusions and recommendations.

Test reports emphasize both fact and opinion. The test data or facts form the actual results and opinions and focus on subjective evaluation of performance. It is important, however, that every conclusion or opinion be substantiated with test data. Table 12.1 provides a recommended test report outline.[1]

Table 12.1: Test Report Contents

- Executive Summary
- Table of Contents
- Section 1
  - Test Background
  - Test Purpose
  - Test Objectives
  - System Configuration and Test Participants
  - Limiting Factors
  - Test Approach
  - Test Control
  - Test Execution
- Section 2
  - Test Results
- Section 3
  - Conclusion
  - Recommendation

The elements listed in Table 12.1 should be viewed as the minimum requirements necessary for an accurate, understandable test report. Although each test is different, all test reports should reflect these requirements. Aside from the suggested minimum requirements, custom-designed sections that explain test performance results might be appropriate.

12.1. References.

1. System Integration Office System, *General Test Report Format* (unclassified and released for publication on August 23, 1990). No date.

Chapter 13.

LESSONS LEARNED

One of the most valuable sources of information available to any testing community is the posttest analysis. Lessons learned is the final step in the final acceptance testing process (see Figure 13.1) but one of the most important. Test managers should use the feedback about lessons learned to determine "what to do" and "what not to do" for future testing. The purpose is to continuously improve the testing process by devising and correcting testing techniques, procedures, and equipment configurations. Even if the same system and configuration is tested continuously, each individual test should be treated as a new system test.

The test manager should ask the following questions when deciding on lessons learned:[1]

1. Were the test time and location appropriate?
2. Was this an appropriate test?
3. Were all test objectives achieved successfully?
4. Did the established test capabilities match the test configuration?
5. Were the test plan contents ample and clear?
6. Are there any corrections to the test procedures?
7. Did the test plan and procedures establish the required guidelines?
8. Were there any problems with test execution?
9. Were there any configuration problems?
10. Were there any timing problems?
11. Were there any problems with data collection equipment or the test data itself? If so, are the test data recoverable?

The following questions provide a lessons learned review for data analysis and test report generation:[2]

1. Are the test data readable and usable?
2. Was the correct time period recorded?
3. Are the correct analysis techniques being utilized?
4. Can the finds be properly substantiated?
5. Are all lessons learned documented in the test report?
6. Have all lessons learned been coordinated with test participants and considered for implementation for future testing of a similar or different system?
7. Did the recipients of the test report agree with its content and format?

Since each test is unique, testers who are members of professional organizations should consider disseminating their lessons learned information to promote the exchange of information. By applying lessons learned we can significantly increase the testing quality and reduce test time and developmental costs. It is appropriate for lessons learned to be included in the test report and presented with discrepancies during staff meetings.

13.1. References.

1. The author's collection of "lessons learned" questions developed from testing experience. Approved for publication.

2. Ibid.

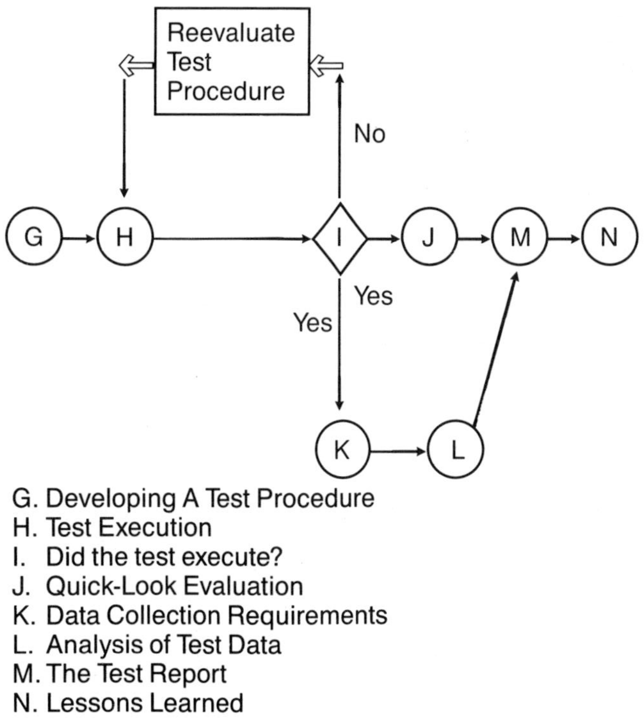

Figure 13.1: *Final Acceptance Testing: Test Execution and Posttest Analysis.* Indicates the author's method and a standard testing sequence of action items during and after test execution. Approved for publication.

Chapter 14.

CONCLUSION AND OBSERVATIONS

Final acceptance testing is a valuable and simple test method available for determining overall system performance as well as user acceptance. In addition to measuring performance, final acceptance testing reports on the system life cycle and measures the capabilities of a new or changed system in its proper operating environment.

Aside from the developer and tester, the final acceptance testing process reveals many benefits to all levels of organizational management, as shown in Table 13.1.[1]

Table 14.1: Benefits of Final Acceptance Testing

- Assess technical integrity
- Complete system test
- Characterize system performance
- Provide operator feedback
- Determine system limitations
- Prove operational concepts
- Benchmark future tests

As discussed in chapters 2 through 13, final acceptance testing from a managerial aspect involves: identifying test requirements and purpose, designing a test, knowing test capabilities, determining test objectives, developing the test plan, developing the test procedures, conducting the actual test, publishing the quick-look test report, data collection, publishing the test report, analyzing test data, and publishing lessons learned.

As a managerial tool, final acceptance testing can be implemented and executed at any management level. It's a simple method of development, implementation, and execution that can be applied by managers over a large spectrum of systems.

14.1. References.

1. Author's personal testing experience and knowledge while assigned to the System Integration Office (SIO).

Appendix 1.

TEST PREPARATION CHECKLIST

Purpose: This checklist serves as a managerial tool for pretest, execution, and posttest data analysis planning requirements.

Directions:

(a) Complete all applicable steps.

(b) This checklist should be utilized with the following
- Data Collection Checklist
- Quick-Look Report Format
- Test Report Questionnaire
- Problem Report

(c) Checklist is concluded when all pretest, execution, and posttest analysis requirements are complete.

General Test Information

Test Name: ____________________ Test Director: ____________________

Test Date: ____________________ Test Location: ____________________

System Identification: ____

Action Items

Accomplish the following

1. Determine test purpose ________
2. Determine critical test issues ________
3. Determine evaluation criteria ________
4. Determine data recording requirements ________
5. Determine system specification ________
6. Determine equipment and system interface ________
7. Determine actual test environment ________
   - on-line system
   - off-line system
8. Determine test objectives ________
9. Determine test stimulus ________
10. Determine sub-objectives ________
11. Determine measurements of performance ________
12. Determine test configuration ________
13. Determine test capabilities ________

Test Design Consideration

14. Simplicity of design _______
15. Sources of variability _______
16. Avoid controlled environments _______
17. Ensure proper test environment _______
18. Variables affecting test outcome _______
19. Dependent variables _______
20. Independent variables _______
21. Test control factors _______

Develop/Review Test Plan and Procedures

22. Required sections for test plan _______
23. Specific steps for test procedure _______
24. Review plans and procedures with all test participants (serves as notification) _______
25. Distribute test plans and procedures prior to the scheduled test date _______

Test Execution Actions

26. Schedule pretest meeting _______
27. Verify that all test participants received test plans and procedures _______
28. Verify all test participants understand test plans and procedures _______
29. Verify test participants understand test purpose _______
30. Schedule posttest briefings _______
31. Verify test participants are aware of: _______
    -data collection requirements
    -problem report format

Quick-Look Report

32. Remind all test participants of report requirement _______
33. Have all test participants document real-time test observations and possible problem areas _______
34. Determine test execution success or failure criteria _______

Data Collection Actions

35. Prepare data collection forms for the test _______
36. Gather all test data during and after the test _______
37. Make sure all participants understand data collection requirements _______
38. Ensure test directors or managers understand requirements _______

Data Analysis

39. Review test data for content _______
40. Perform analysis _______
41. Determine test discrepancies _______
42. Document and report test findings _______
43. Assess system performance _______

Test Report

44. Review all test inputs _______
45. Organize test report contents _______
46. Review Test Report Checklist _______
47. Determine test report distribution _______

Lessons Learned

48. Review entire test process to determine overall test improvements ______
49. Identify system improvements ______
50. Identify human improvements ______
51. Document for future test designs ______

Designed by Michael J. Muolo

Appendix 2.

TEST PARTICIPANT NOTIFICATION

Test Name: ____________	Report Date: ____________
Test Location: ____________	Test Director: ____________
Test Date: ____________	System Identified: ____________
Phone: ____________	Firm: ____________

Directions:
(a) Participant: Indicate name of test participant.
(b) Location: Test participant organization/agency/firm.
(c) Position: Test participant position during the test.
(d) Phone: Test participant.
(e) Date: Test participant contacted on this day.
(f) Use continuation sheet if required.

PARTICIPANT	LOCATION	POSITION	PHONE	DATE

Designed by Michael J. Muolo

Appendix 3.

DATA COLLECTION PROCEDURE

Test Name: ______________________	Report Date: ______________________
Test Location: ______________________	Tester: ______________________
Test Date: ______________________	Organization: ______________________
Phone: ______________________	System Identified: ______________________

Position: ____

____

Directions: (a) Check off and complete each item. (b) Return all test data to the test director.

1. Data recording equipment on and operational. If data recording equipment cannot be activated immediately at test start, contact the test manager or director. ______

2. If data recording equipment involves a hard copy output, verify printer operability. ______

3. Check to see if ample paper or recording medium is available for entire test. ______

4. Relay recording status to test director. Periodically pass recording status to the test director. ______

5. Record and collect all test data for analysis. ______

6. Record all real-time observations for the Quick-Look report. Reference Quick-Look format ______

____

Remarks (Use reverse side if required):

____

Designed by Michael J. Muolo

Appendix 4.

TEST REPORT CHECKLIST

Purpose: This checklist serves as a managerial tool for the organization and final publication of test report.

Test Name: ______________	Report Date: ______________
Test Location: ______________	Firm: ______________
Test Date: ______________	System Identification: ______________
Phone: ______________	Test Director: ______________

Directions:
(a) Review each item listed and annotate (X) indicating compliance.
(b) Each step corresponds to Table 12.1, "Test Report Contents."
(c) Progressively draft each section as information or test data become available.
(d) Each section contains recommended actions only and aren't all-inclusive. Additional sections can be included to enhance the checklist.

General Test Information

1. Provide an executive summary ______

This section is an overview of the system tested, purpose, outcome, and conclusion. It represents a summary of all test report sections and should provide the bottom line regarding system performance to upper-level management.

2. Ensure a table of contents ______

Section 1

3. Provide a brief test background ______

Include general historical information on how the requirement for this test derived.

4. Include the test purpose ______

5. Include the test objectives ______

Ensure all test objectives and sub-objectives are clearly defined. Include the individual measures of effectiveness for each.

6. Provide a diagram depicting the entire test configuration and identify individual test participants ______

7. Describe individual limiting factors and their effect on test outcome ______

Include technical, operational, environmental, and human consideration

8. Completely describe the mechanics of the test ______

9. Explain in detail the testing configuration ______

Indicate the on-line or off-line testing configuration

10. Comment on the test execution ______

Was it a smooth test? Include all problem areas and work around during the test.

Section 2

11. Include all test results ______

Results should be explained in detail by test objective or sub-objective

Section 3

12. The conclusion must specify test success or failure ______

13. Recommendations should concentrate on lessons learned. ______

Designed by Michael J. Muolo

Appendix 5.

TEST PROCEDURES FOR RDS-120 TRACKING SYSTEM

Test Purpose

The RDS-120 tracking system is used to track aircraft for small independently operated private airports. This test will demonstrate that operational and system technical specifications are suitable and attainable.

Test Procedures Organization

These test procedures are all-inclusive with respect to system configuration, test execution, test data collection requirements, system reconfiguration, and test termination.

General Information

This test will last about six hours, allowing for 10 scheduled aircraft arrivals and five departures. The scheduled test date is January 2 1991, 8 a.m., at Radar-Systems test facility. The secondary test date is January 3 1991. A quick-look test report will be published and distributed on January 8 1991.

Abbreviations

TD	Test Director
DO	Display Operator
RR	Radar Receiver
IPE	Intermediate Processing Equipment
CPU	Central Processing Unit
OD-P	Operational Display-Primary
OD-S	Operational Display-Secondary
RC	Radar Chain
RA	Radar Antenna
FC	Facility Controller
TF	Test Facility
MC	Maintenance Control
TP	Test Participant
SE	System Engineer

TEST PROCEDURES

Step No.	Action	Description	Completed
1.	TD	Notify all test participants 72 hours prior to scheduled date.	Yes No
2.	TD	Direct DO to review test plan and procedures.	Yes No
3.	TD	Verify that participants are ready for test.	Yes No

Step No.	Action	Description	Completed
4.	TD	Schedule a working group meeting to review test requirements.	Yes No
5.	TD	Verify FC and TF are ready 24 hours before test date.	Yes No
6.	FC	Begin TF configuration 12 hours before test date.	Yes No
7.	FC	Verify RA operations	Yes No
8.	FC	Verify RR operations	Yes No
9.	FC	Verify IPE operations	Yes No
10.	FC	Verify CPU operations	Yes No
11.	FC	Verify OD-P operations	Yes No
12.	FC	Verify OD-S operations	Yes No
13.	FC	Verify data recording equipment operations	Yes No
14.	FC	Report system discrepancies to MC for repair if required	Yes No
15.	FC	Report test readiness to TD immediately after equipment checkout (steps 7-13)	Yes No
16.	TD	Verify test date with FC	Yes No
17.	TD/FC	Conduct pre-brief of test requirements with DO	Yes No
18.	TD	Ensure all TP's are aware of data collection requirements	Yes No
19.	TD/FC	Execute test	Yes No

Step No.	Action	Description	Completed
20.	TD/FC	Reschedule test if required and reinstitute test procedures.	Yes No
21.	TD	Direct DO to track all aircraft arrivals.	Yes No
		Arrival 1	Time ______
		Arrival 2	Time ______
		Arrival 3	Time ______
		Arrival 4	Time ______
		Arrival 5	Time ______
		Arrival 6	Time ______
		Arrival 7	Time ______
		Arrival 8	Time ______
		Arrival 9	Time ______
		Arrival 10	Time ______
22.	TD	Direct DO to track all aircraft departures	Yes No
		Departure 1	Time ______
		Departure 2	Time ______
		Departure 3	Time ______
		Departure 4	Time ______
		Departure 5	Time ______
23.	TD	Verify operations of OD - P and OD - S	Yes No
24.	TD	Collect all recorded data for analysis (reference data collection checklist)	Yes No
25.	FC	Terminate test	Time ______
26.	TD	Deactivate radar system according to maintenance procedures	Yes No

Step No.	Action	Description	Completed
27.	TD	Conduct posttest briefing with all test participants to discuss initial test findings	Yes No
28.	TD	Obtain operator feedback on system usability	Yes No
29.	TD	Publish quick-look report immediately following scheduled test date	Yes No
30.	TD	Direct SE to perform data analysis	Yes No
31.	TD	Review test data and determine usability of system	Yes No
32.	TD	Publish final test report	Yes No
33.	TD	Publish lessons learned findings	Yes No

Appendix 6.

PROBLEM REPORT

Test Name: ______________	Report Date: ______________
Test Location: ______________	Firm: ______________
Test Date: ______________	System Identification: ______________
Phone: ______________	Test Director: ______________

Directions:
(a) Complete all applicable areas.
(b) Be as specific as possible in identifying the problem.
(c) Use the reverse side and or additional pages if necessary.
(d) Return the completed form along with test data to the test director.
(e) Identify one problem per form.

---

1. This report addresses the following system components: (X appropriate)

Hardware___ Man/Machine Interface___ Test Configuration___
Software___ Test Plan___ Test Execution___
Firmware___ Test Procedures___ Test Termination___
Equipment Configuration___ Test Data___
Other (specify below) ___

Other______________________________

____

To Be Completed By The Test Participant

2. Complete and or circle the following:
   a. Explain the problem area: ____________________

   ____

   b. Indicate the level of concern: High Med Low
   c. Did problem(s) affect test outcome? Yes No
   d. When was the problem discovered: ____________________

   ____

   e. Any test data available or collected? Yes No
   If yes, attach to this report and explain at f below
   f. Identify test data collected: ____________________

   ____

g. Has any troubleshooting commenced? Yes No
If yes, discuss at h below

h. Identify troubleshooting to date: ______________________________

____

i. Has corrective action commenced or completed? Yes No
If yes, identify at j below

j. Indicate corrective action: ______________________________

____

k. This problem was identified to whom? ______________________
Indicate agency, individual, or firm

To Be Completed By Maintenance

l. Can the system function with this problem? Yes No
Explain ______________________________________

____

m. Are enough test data available to
troubleshoot and/or correct this problem? Yes No
If yes, indicate at n below

n. Indicate required test data: ______________________________

____

o. Projected data for correction: ____________________________

p. Is outside expertise required to correct this problem? Yes No
If yes, indicate at q below

q. Identify consulting expertise required and cost: ______________

____

____

3. If this problem is procedurally orientated, identify recommended actions:

____

____

Designed by Michael J. Muolo

Appendix 7.

QUICK-LOOK TEST REPORT

Test Name: ____________________ Report Date: ____________________

Test Location: ____________________ Test Director: ____________________

Test Date: ____________________ System Identified: ____________________

Phone: ____________________ Firm: ____________________

Position: ____

Directions:
(a) Answer or circle the following.
(b) Be as specific as possible in identifying initial test results.
(c) Use the reverse side or additional paper if necessary.
(d) Return the completed form to the test director.

---

1. Complete the following:

   a. Test start time: ____________________

   b. Test termination time: ____________________

   c. Were test objectives satisfied? Yes No
      If no, explain: ____________________

   d. Did your equipment operate to test completion? Yes No
      If no, explain: ____________________

   e. Were all test events observed? Yes No
      If no, explain: ____________________

   f. Was operational display quality acceptable? Yes No
      If no, explain: ____________________

   g. Were there any software problems noted? Yes No
      If yes, explain: ____________________

   h. Were there any hardware problems noted? Yes No
      If yes, explain: ____________________

   i. Were test plan requirements clear and concise? Yes No
      If no, explain: ____________________

   j. Were test procedures complete and accurate? Yes No
      If no, explain: ____________________

   k. Was the test design understandable?
      Did it satisfy test objectives? Yes No
      If no, explain: ____________________

l. Any problems noted with data collection? Yes No
(identify in data collection checklist)

m. Did the system react as specified or designed? Yes No
If no, explain: ______________________________

n. Any problems noted with the system's human/machine interface? Yes No
If no, explain: ______________________________

o. Does the system satisfy operational requirements? Yes No
If no, explain: ______________________________

2. Additional comments: ______________________________

____

____

____

____

____

Designed by Michael J. Muolo

Bibliography

Department of the Air Force. *Air Force Regulation 55-43, Management of Operational Test and Evaluation.* Air Force Operational Test and Evaluation Center, Kirtland Air Force Base, New Mexico. June 28, 1985.

Director of Analysis, Air Force Operational Test and Evaluation Center, *Technical Paper 3.0, Test Design.* Kirtland Air Force Base, New Mexico. May 1986.

General System Integration Office. *Quick-Look Report Format.* No date.

Stevens, Roger T. *Operational Test & Evaluation: A Systems Engineering Process.* Malabar, Florida: Robert E. Krieger Publishing Company, 1986.

System Integration Office (SIO). *Ballistic Missile Tactical Warning and Attack Assessment System High Load Test Procedures.* October 9, 1990.

———. *Combined Ballistic Missile and Atmospheric TW/AA System End-to-End Performance Test Procedures.* October 9, 1989.

———. *"Data Collection Requirements Checklist" to support End-to-End Performance Testing.* No date.

———. Directorate of Test and Certification. *SIO Standard 3000: Test and Demonstration Standard.* May 30, 1986.

———. Directorate of Test and Certification. *SIO Standard 3100: Test Capability Standard.* November 29, 1985.

———. *System Test Report Format for General High Load System Testing.* No date.

Note: All documents, figures, and tables are unclassified and released for reference and use in this publication.

Index

Printed on Recycled Paper